SEEING

by Robin Nelson

first step nonfiction

Lerner Publications Company · Minneapolis

Seeing is one of my **senses.**

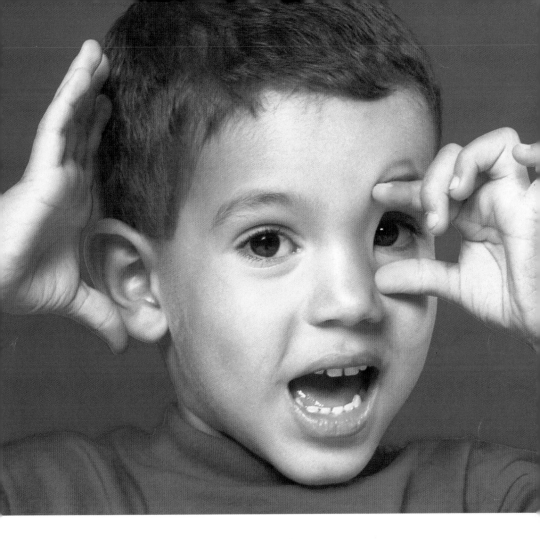

I see with my eyes.

I see colors.

I see a blue sky.

I see a red barn.

I see a yellow ball.

I see shapes. I see a **circle.**

I see a **square.**

I see a **triangle.**

I see a star.

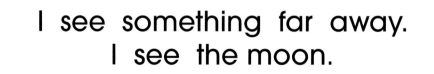

I see something far away.
I see the moon.

I see a **mountain.**

I see something close up.
I see fish.

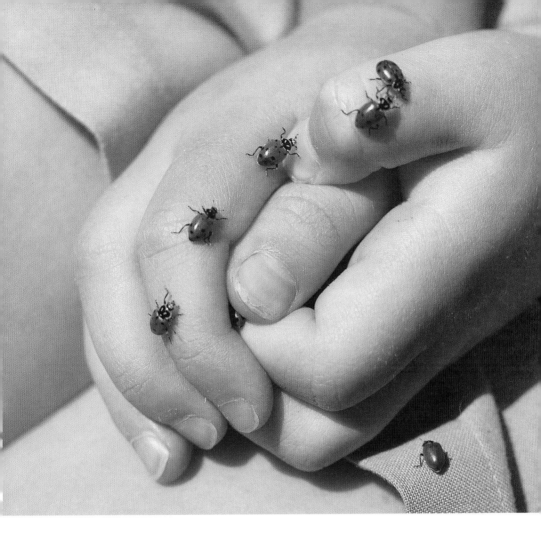

I see bugs.

I see many things.

What do you see?

iris

pupil　　　lens

How do you see?

When you see something, your eye lets in light through the pupil. The lens of your eye makes a picture at the back of your eye. The picture is upside down. Your brain turns the picture right side up for you. Your brain tells you what you are seeing. The colored part of your eye is called the iris. The iris helps to let in the right amount of light.

Seeing Facts

 A frog can see almost all the way around its body without turning its head. A frog has eyes on top of its head. A frog turns its eyes around to look behind it.

 Birds can't move their eyes. They have to turn their head to look around.

An ostrich's eyes are two inches across. Each eye weighs more than the animal's brain.

A chameleon's eyes can look in opposite directions at the same time.

 Your eyes stop growing when you are about seven years old.

 A newborn baby sees the world upside down. It takes some time for the baby's brain to learn to turn the picture right-side up.

 The giant squid has the largest eyes of any creature.

Glossary

 circle – a perfectly round shape with no corners

 mountain – a high piece of land

 senses – the five ways our bodies get information. The five senses are hearing, seeing, smelling, tasting, and touching.

 square – a shape that has four sides of the same length

 triangle – a shape that has three straight sides

Index

close up – 14, 15

colors – 4, 5, 6, 7

eyes – 3, 18–19

far away – 12, 13

shapes – 8, 9, 10, 11

Cover image used courtesy of: Brand X Pictures.

Photos reproduced with the permission of: © Michael Pole/CORBIS, pp. 2, 22 (middle); © Darwin Wiggett/CORBIS, p. 3; © RubberBall Royalty Free Digital Stock Photography, p. 4; © Darrell Gulin/CORBIS, p. 5; © Richard Cummins, p. 6; Brand X Pictures, pp. 7, 17; © Stephen G. Donaldson, pp. 8, 22 (top); © Ludovic Maisant/CORBIS, pp. 9, 22 (second from top); © Larry Neubauer/CORBIS, pp. 10, 22 (bottom); Corbis Royalty Free Images, p. 11; © Betty Crowell, pp. 12, 15; © Richard B. Levine, pp. 13, 22 (second from top); © Photo Network, p. 14; © Doug Crouch/CORBIS, p. 16.

Illustration on page 18 by Tim Seeley.

Lerner Publications Company
A division of Lerner Publishing Group
241 First Avenue North
Minneapolis, MN 55401 U.S.A.

Website address: www.lernerbooks.com

Library of Congress Cataloging-in-Publication Data

Nelson, Robin, 1971–
    Seeing / by Robin Nelson.
      p.   cm. — (First step nonfiction)
    Includes index.
    Summary: An introduction to the sense of sight and the different things that you can see.
    ISBN: 0–8225–1262–9 (lib. bdg. : alk. paper)
    1. Vision—Juvenile literature. [1. Vision. 2. Eye. 3. Senses and sensation.] I. Title.
    II. Series.
QP475.7.N45 2002
612.8'4—dc21   2001003966

Manufactured in the United States of America
2 3 4 5 6 7 – DP – 08 07 06 05 04 03